15 TELESCOPES THAT CHANGED OUR VIEW OF THE UNIVERSE

By Rayan Bale

Your Gateway to the Stars

Copyright © 2024 by Rayan Bale

All rights reserved. No part of this publication may be reproduced, distributed, or transmitted in any form or by any means, including photocopying, recording, or other electronic or mechanical methods, without the prior written permission of the publisher, except in the case of brief quotations embodied in critical reviews and certain other noncommercial uses permitted by copyright law.

TABLE OF CONTENTS

Introduction — 3

Telescope 1: Galileo's Telescope (1609): The First Telescope — 7

Telescope 2: Yerkes 40-inch Refractor (1897): The Largest Refractor — 12

Telescope 3: Mount Wilson 60-inch Reflector (1908): Pioneering Reflector — 18

Telescope 4: Hooker 100-inch Telescope (1917): The Expanding Universe — 24

Telescope 5: Hale 200-inch Telescope (1948): The Giant Leap — 30

Telescope 6: Horn Antenna (1959): Echoes of the Big Bang — 36

Telescope 7: Arecibo Observatory (1963): The Giant Ear — 42

Telescope 8: Westerbork Synthesis Radio Telescope (1970): Radio Interferometry Pioneer — 48

Telescope 9: Very Large Array (1980): **54**
    Radio Visionaries

Telescope 10: Hubble Space Telescope **60** (1990): Eyes in the Sky

Telescope 11: Keck 1 & 2 Telescopes **66** (1993): The Twin Giants

Telescope 12: Chandra X-ray **72** Observatory (1999): X-ray Vision

Telescope 13: Spitzer Space Telescope **78** (2003): Infrared Explorer

Telescope 14: Atacama Large Millimeter **84** Array (ALMA, 2011): The Cosmic Eye

Telescope 15: James Webb Space **90** Telescope (2021): A New Vision

Conclusion **96**

References **100**

INTRODUCTION

The night sky has always been a source of wonder and curiosity for humanity. From the earliest days of stargazing, people have sought to understand the cosmos and our place within it. Over the centuries, the development of telescopes has revolutionized our view of the universe, revealing its vastness, complexity, and beauty.

This book, "15 Telescopes That Changed Our View of the Universe," takes you on a journey through the history of astronomy, highlighting the most significant telescopes and their contributions to our understanding of the cosmos. Each chapter delves into the technological innovations, groundbreaking discoveries, and lasting legacies of these remarkable instruments.

We begin with Galileo's Telescope, the first practical telescope that opened humanity's eyes to the moons of Jupiter and the phases of Venus, challenging the geocentric view of the universe. From there, we move through the centuries, exploring the advances in telescope design and the profound insights they have provided.

The Yerkes 40-inch Refractor and the Mount Wilson 60-inch Reflector marked significant leaps in our ability to observe the stars, while the Hooker 100-inch Telescope revealed the expanding universe and the vastness of the cosmos. The Hale 200-inch Telescope continued this tradition, pushing the boundaries of optical astronomy.

With the advent of radio and space-based telescopes, our understanding of the universe expanded even further. The Horn Antenna detected the echoes of the Big Bang, and the Arecibo Observatory became a giant ear, listening to the whispers of the cosmos. The Westerbork Synthesis Radio Telescope pioneered radio interferometry, and the Very Large Array provided radio visionaries with detailed views of the universe.

The Hubble Space Telescope, with its eyes in the sky, brought the wonders of the cosmos into stunning clarity, capturing images that have captivated the world. The Keck 1 & 2 Telescopes used their twin giants to explore distant galaxies and exoplanets, while the Chandra X-ray Observatory offered X-ray vision into the high-energy universe.

The Spitzer Space Telescope explored the infrared universe, revealing the hidden processes of star and planet formation. The Atacama Large Millimeter

Array (ALMA) provided a cosmic eye into the cold and dusty regions of space, and the James Webb Space Telescope ushered in a new vision for space-based astronomy, promising to transform our understanding of the early universe, exoplanets, and galactic evolution.

Each of these telescopes represents a milestone in our quest to explore the cosmos. They have expanded our horizons, challenged our perceptions, and deepened our understanding of the universe. As we look to the future, these instruments and the discoveries they have enabled continue to inspire new generations of astronomers and scientists.

Join us on this journey through the history of telescopes and the wonders they have revealed. Discover the stories behind these remarkable instruments and the scientists who used them to unlock the secrets of the universe. This is the story of our exploration of the cosmos, told through the eyes of the telescopes that changed our view of the universe.

TELESCOPE I:
GALILEO'S TELESCOPE (1609)

THE FIRST TELESCOPE

Introduction of Galileo's Telescope:

In 1609, Galileo Galilei revolutionized astronomy with the creation of the first practical telescope. This groundbreaking instrument allowed Galileo to make unprecedented observations of the night sky, forever altering our understanding of the universe and our place within it.

The Construction of the Telescope:

Galileo's telescope had a simple yet innovative design. It consisted of a convex objective lens and a concave eyepiece, mounted in a wooden tube. Despite its modest magnification of about 20x, the telescope was powerful enough to reveal previously unseen details of the Moon's surface and the phases of Venus. This basic but effective design paved the way for future advancements in telescope technology, laying the foundation for modern observational astronomy.

Discoveries Through the Telescope:

The true power of Galileo's telescope lay in its ability to unveil the secrets of the cosmos. Among his most significant discoveries were:

- **Jupiter's Moons:** Galileo observed four large moons orbiting Jupiter—Io, Europa, Ganymede, and Callisto—now known as the Galilean moons. This discovery provided the first clear evidence that not all celestial bodies orbited the Earth, challenging the geocentric model of the universe. It was a revolutionary finding that suggested a more complex and dynamic solar system than previously imagined.

- **Phases of Venus:** By observing Venus, Galileo noted that it exhibited phases similar to those of the Moon, proving that Venus orbited the Sun. This observation was a crucial piece of evidence supporting the heliocentric model proposed by Copernicus, which posited that the planets, including Earth, revolved around the Sun. Galileo's findings provided strong support for this controversial theory.

- **Surface of the Moon:** Contrary to the belief that the Moon was a perfect, smooth sphere, Galileo discovered that it had mountains, valleys, and craters, much like the Earth. This observation challenged the Aristotelian view of celestial perfection and suggested that celestial bodies were more similar to Earth than previously thought.

The Legacy of Galileo's Telescope:

Galileo's observations, published in his 1610 book Sidereus Nuncius (Starry Messenger), shook the foundations of the scientific and religious communities. The evidence he provided supported the Copernican theory that the Earth and other planets orbited the Sun. This heliocentric view contradicted the geocentric model endorsed by the Catholic Church, leading to significant controversy and conflict.

Despite facing significant opposition from the Catholic Church, which held firm to the geocentric model, Galileo's work laid the groundwork for modern astronomy. His discoveries not only advanced scientific knowledge but also encouraged a spirit of inquiry and skepticism that is central to the scientific method. Galileo's trial by the Inquisition, house arrest, and eventual forced recantation are well-documented. However, in 1992, more than 350 years after his death, the Catholic Church formally acknowledged its error in condemning Galileo.

NASA honored Galileo's legacy by naming a spacecraft after him; the Galileo mission, which studied Jupiter and its moons, paying tribute to the scientist's pioneering discoveries.

Galileo's telescope marked the beginning of observational astronomy, transforming it from a theoretical to an empirical science. His work inspired future generations of scientists, including Johannes Kepler and Isaac Newton, who built upon his discoveries to develop their own groundbreaking theories. Galileo's approach to scientific observation and his reliance on empirical evidence set a standard that continues to influence scientific inquiry to this day.

Galileo's telescope was more than just an instrument; it was a catalyst for a scientific revolution. By looking through his simple yet powerful device, Galileo opened humanity's eyes to the vast and complex universe beyond our planet. His legacy continues to inspire astronomers and scientists to this day, reminding us of the power of observation and the endless possibilities of discovery. The eventual pardon by the Church serves as a testament to the enduring truth of his findings and the importance of perseverance in the pursuit of knowledge.

TELESCOPE 2:
YERKES 40-INCH REFRACTOR (1897)
THE LARGEST REFRACTOR

Introduction of the Yerkes Refractor:

In 1897, the Yerkes Observatory in Williams Bay, Wisconsin, became home to the largest refracting telescope ever built, the Yerkes 40-inch Refractor. This monumental achievement in telescope construction marked a significant milestone in astronomical observation, enabling scientists to explore the universe with unprecedented clarity and detail.

The Construction of the Yerkes Refractor:

The Yerkes Refractor was an engineering marvel of its time. Designed by the renowned optician Alvan Clark and his sons, the telescope boasted a 40-inch diameter lens, the largest ever crafted for a refracting telescope. The lens was mounted in a 63-foot-long tube, supported by a massive iron structure that allowed for precise adjustments and tracking of celestial objects. The telescope's mount, a masterpiece of mechanical engineering, enabled smooth and accurate movements, crucial for long exposure astrophotography and detailed observations.

The construction of the Yerkes Refractor involved significant challenges and innovations. The massive lens, made from high-quality glass, required precise grinding and polishing to achieve the necessary optical clarity. The support structure, designed to hold the heavy lens and tube, needed to be robust yet finely balanced to prevent any vibrations or distortions during observations. This combination of optical precision and mechanical stability made the Yerkes Refractor a state-of-the-art instrument for its time.

Discoveries and Contributions:

The Yerkes Refractor quickly became a powerful tool for astronomers, contributing to numerous discoveries and advancements in the field:

- **Stellar Photography:** The large aperture of the Yerkes Refractor allowed astronomers to capture detailed photographs of stars, nebulae, and other celestial objects. These images provided valuable data for studying the structure and composition of the universe. The clarity and detail of these photographs were unmatched by earlier instruments, enabling new insights into the nature of the cosmos.

- **Spectroscopy:** The telescope's high-resolution optics made it ideal for spectroscopic studies, enabling scientists to analyze the light emitted by stars and other celestial bodies. This led to significant advancements in our understanding of stellar composition, temperature, and motion. By examining the spectra of various stars, astronomers could determine their chemical compositions, velocities, and even the presence of companion stars or planets.

- **Double Stars and Star Clusters:** The Yerkes Refractor was instrumental in the study of double stars and star clusters. Astronomers used the telescope to measure the positions and motions of these objects, providing insights into the dynamics of stellar systems. These observations helped to refine models of stellar evolution and interactions, shedding light on the life cycles of stars and the formation of complex systems.

- **Planetary Studies:** The Yerkes Refractor was also used to observe planets within our solar system, providing detailed images and data that improved our understanding of their atmospheres, surfaces, and moons. These observations were critical for mapping the features of planets like Mars and Jupiter and studying phenomena such as the Great Red Spot and planetary rings.

Educational and Public Outreach:

In addition to its scientific contributions, the Yerkes Refractor played a vital role in education and public outreach. The observatory became a hub for training the next generation of astronomers, offering hands-on experience with cutting-edge technology. Many notable astronomers, including Edwin Hubble, began their careers at Yerkes, learning the skills and techniques that would later lead to groundbreaking discoveries.

The Yerkes Observatory also welcomed visitors from the public, providing a unique opportunity to view the heavens through a world-class telescope. Public observing nights and educational programs helped to foster a broader appreciation for astronomy and science in general. The impressive size and capabilities of the Yerkes Refractor captivated the imagination of countless visitors, inspiring future scientists and enthusiasts.

The Legacy of the Yerkes Refractor:

The Yerkes 40-inch Refractor remained the largest operational refracting telescope in the world and held this title well into the 20th century. Its contributions to astronomy were immense, shaping the course of astronomical research and education.

The telescope became a training ground for many prominent astronomers, including Edwin Hubble, who later made groundbreaking discoveries with the Hooker 100-inch Telescope at Mount Wilson Observatory.

Despite advancements in reflecting telescope technology, the Yerkes Refractor continued to be a valuable tool for education and outreach. Its impressive size and historical significance made it a popular attraction for visitors and a symbol of the pioneering spirit of astronomical exploration. The observatory's commitment to public engagement ensured that the legacy of the Yerkes Refractor would endure, inspiring new generations to explore the mysteries of the universe.

The Yerkes 40-inch Refractor stands as a testament to the ingenuity and ambition of late 19th-century astronomers and engineers. Its construction marked a pinnacle in the development of refracting telescopes, providing a foundation for future advancements in observational astronomy. The legacy of the Yerkes Refractor lives on, inspiring new generations of astronomers and reminding us of the enduring quest to understand the cosmos. The observatory and its iconic telescope continue to symbolize the relentless pursuit of knowledge and the wonders that await discovery in the vast expanse of space.

TELESCOPE 3: MOUNT WILSON 60-INCH REFLECTOR (1908)

PIONEERING REFLECTOR

Introduction of the Mount Wilson 60-inch Reflector:

In 1908, the Mount Wilson Observatory in California unveiled the 60-inch Reflector, a groundbreaking instrument that heralded a new era in astronomical observation. This telescope, designed by George Ellery Hale, was the largest operational reflecting telescope at the time and set the stage for numerous astronomical discoveries that would redefine our understanding of the universe.

The Construction of the 60-inch Reflector:

The Mount Wilson 60-inch Reflector was an engineering marvel. Its primary mirror, made of glass and coated with a reflective layer of silver, had a diameter of 60 inches, making it the largest reflecting telescope in the world at the time. The mirror was mounted in a sturdy steel frame, allowing for precise adjustments and stable observations.

The construction process involved significant challenges, particularly in crafting the large, perfectly shaped mirror and developing a support structure capable of handling its weight and maintaining its alignment. The telescope's innovative design included a Coudé focus system, which allowed light to be directed to various instruments, facilitating detailed analysis of celestial objects.

Discoveries and Contributions:

The 60-inch Reflector quickly became one of the most productive telescopes of its era, contributing to a multitude of discoveries and advancements in astronomy:

- **Cepheid Variables and the Distance Scale:** One of the most significant contributions of the 60-inch Reflector was the study of Cepheid variable stars. Astronomer Harlow Shapley used these stars to measure distances to far-off star clusters, leading to a more accurate understanding of the size of our galaxy and the distribution of stars within it. This work laid the groundwork for the concept of the universe's vast scale.

- **Structure of the Milky Way:** The telescope's powerful optics allowed astronomers to resolve individual stars in distant star clusters and nebulae, leading to a better understanding of the structure and composition of the Milky Way. Observations made with the 60-inch Reflector helped to reveal the spiral nature of our galaxy and the presence of star-forming regions.

- **Stellar Spectroscopy:** The 60-inch Reflector was also instrumental in advancing the field of stellar spectroscopy. By analyzing the light from stars, astronomers were able to determine their chemical compositions, temperatures, and velocities. This information provided insights into stellar evolution and the lifecycle of stars.

Technological Innovations:

The Mount Wilson 60-inch Reflector introduced several technological innovations that would influence the design of future telescopes. The use of a large, silver-coated glass mirror allowed for greater light-gathering power and sharper images than earlier telescopes. The Coudé focus system enabled versatile use of the telescope with different scientific instruments, enhancing its capability for detailed analysis.

Furthermore, the telescope's mount and drive systems were designed to provide smooth, precise tracking of celestial objects, compensating for the Earth's rotation. This precision was crucial for long exposure astrophotography and accurate measurements, making the 60-inch Reflector a versatile and powerful tool for astronomers.

Educational and Public Impact:

Beyond its scientific contributions, the Mount Wilson 60-inch Reflector played a vital role in education and public outreach. The observatory became a center for astronomical research and education, attracting prominent astronomers from around the world. It provided training and experience for many who would go on to make significant contributions to astronomy.

The observatory also opened its doors to the public, offering opportunities for people to view celestial objects through the powerful telescope. Public observing nights and educational programs helped to spark interest in astronomy and inspire future generations of scientists.

The Legacy of the Mount Wilson 60-inch Reflector:

The Mount Wilson 60-inch Reflector remained the largest telescope of its kind until it was surpassed by the 100-inch Hooker Telescope, also at Mount Wilson Observatory. Despite being eclipsed in size, the 60-inch Reflector continued to be a valuable instrument for astronomical research and education for many decades.

Its legacy is marked by the numerous discoveries it facilitated and the technological advancements it introduced. The telescope set new standards for optical and mechanical design in reflecting telescopes, influencing the construction of future instruments. It also demonstrated the value of large, ground-based telescopes for advancing our understanding of the universe.

The Mount Wilson 60-inch Reflector stands as a testament to the pioneering spirit of early 20th-century astronomers and engineers. Its construction marked a significant advancement in telescope technology, enabling new discoveries and expanding our knowledge of the cosmos. The legacy of the 60-inch Reflector lives on, inspiring astronomers and engineers to continue pushing the boundaries of what is possible in the quest to explore the universe.

TELESCOPE 4: HOOKER 100-INCH TELESCOPE (1917)

THE EXPANDING UNIVERSE

Introduction of the Hooker 100-inch Telescope:

In 1917, the Mount Wilson Observatory in California became home to the Hooker 100-inch Telescope, the largest telescope in the world at that time. Designed by George Ellery Hale, this monumental instrument played a pivotal role in some of the most significant discoveries in 20th-century astronomy, including the realization that the universe is expanding.

The Construction of the Hooker Telescope:

The Hooker 100-inch Telescope was an engineering marvel, featuring a primary mirror 100 inches in diameter. The mirror, crafted from a single piece of glass, was meticulously ground and polished to achieve the precise curvature needed for clear, sharp images. The telescope's tube and mounting structure, made of steel, were designed to support the massive mirror and allow for precise movements and tracking of celestial objects.

Constructing the Hooker Telescope involved overcoming numerous technical challenges, particularly in the production and handling of the large mirror. The entire assembly was housed in a large dome, designed to protect the telescope and provide a stable environment for observations. This combination of advanced optical and mechanical design made the Hooker Telescope a state-of-the-art instrument for its time.

Discoveries and Contributions:

The Hooker 100-inch Telescope quickly became a cornerstone of astronomical research, contributing to several groundbreaking discoveries:

- **The Expanding Universe:** Perhaps the most famous discovery made with the Hooker Telescope was Edwin Hubble's observation that the universe is expanding. By studying the redshifts of distant galaxies, Hubble determined that they were moving away from us, implying that the universe was expanding. This discovery provided key evidence for the Big Bang theory and fundamentally changed our understanding of the cosmos.

- **Cepheid Variables and Galactic Distances:** Building on earlier work with the 60-inch Reflector, Hubble used the Hooker Telescope to observe Cepheid variable stars in the Andromeda Galaxy. These observations allowed him to measure the distance to Andromeda, proving that it was a separate galaxy far beyond the Milky

Way. This revelation expanded the known scale of the universe and established the existence of multiple galaxies.

- **Dark Matter Evidence:** The Hooker Telescope was also instrumental in the discovery of evidence for dark matter. Astronomer Fritz Zwicky used the telescope to study the Coma Cluster of galaxies, finding that the visible mass was insufficient to account for the observed gravitational effects. This led to the hypothesis of dark matter, an unseen substance that makes up much of the universe's mass.

Technological Innovations:

The Hooker Telescope introduced several technological advancements that set new standards for astronomical instruments. The large, precise mirror provided unparalleled light-gathering power, allowing astronomers to observe faint and distant objects. The telescope's mount and drive systems were designed for smooth, accurate tracking, essential for long exposure astrophotography and detailed measurements.

One of the key innovations was the use of a siderostat, a flat mirror that redirected light from celestial objects into the telescope. This allowed for continuous observation of an object as it moved across the sky, greatly enhancing the telescope's capabilities. The Hooker Telescope's success demonstrated the feasibility and benefits of large-aperture reflecting telescopes, influencing the design of future instruments.

Educational and Public Impact:

The Hooker 100-inch Telescope not only advanced scientific knowledge but also played a significant role in education and public outreach. The Mount Wilson Observatory became a hub for astronomical research and training, attracting prominent scientists from around the world. It provided a platform for collaborative research and fostered the development of new techniques and technologies in astronomy.

Public interest in the Hooker Telescope was high, and the observatory hosted numerous public viewing nights and educational programs. These events allowed people to experience the wonders of the universe firsthand and inspired many to pursue careers in science and astronomy. The telescope's achievements were widely publicized, helping to raise public awareness and appreciation of astronomical research.

The Legacy of the Hooker 100-inch Telescope:

The Hooker 100-inch Telescope remained the largest operational telescope in the world until it was surpassed by the Hale 200-inch Telescope at Palomar Observatory. Despite being eclipsed in size, the Hooker Telescope continued to be a valuable instrument for research and education for many years.

Its legacy is marked by the numerous discoveries it facilitated and the technological innovations it introduced. The Hooker Telescope played a crucial role in expanding our understanding of the universe and set new standards for the design and construction of large telescopes. It also demonstrated the importance of collaboration and innovation in scientific research, paving the way for future advancements in astronomy.

The Hooker 100-inch Telescope revolutionized astronomy with its groundbreaking discoveries and advanced technology. Its legacy continues to inspire astronomers, emphasizing the importance of innovation and perseverance in exploring the universe.

TELESCOPE 5:
HALE 200-INCH TELESCOPE (1948)

THE GIANT LEAP

Introduction of the Hale 200-inch Telescope:

In 1948, the Palomar Observatory in California unveiled the Hale 200-inch Telescope, named in honor of George Ellery Hale. This telescope, the largest optical telescope in the world at its time, represented a significant leap forward in astronomical observation and technology. It remained the largest for over four decades and played a crucial role in many groundbreaking discoveries.

The Construction of the Hale Telescope:

The Hale Telescope's construction was an ambitious and complex project that spanned nearly two decades. The primary mirror, 200 inches (5.1 meters) in diameter, was cast from a new type of glass called Pyrex, which had a low thermal expansion coefficient, reducing distortion due to temperature changes. The mirror was meticulously ground and polished to achieve the precise curvature required for clear, detailed observations.

The telescope's mount and dome were equally impressive, designed to support and precisely move the massive mirror. The mount allowed for smooth tracking of celestial objects, essential for long exposure photography and accurate data collection. The dome, a marvel of engineering, protected the telescope while allowing it to observe the night sky without obstruction.

Discoveries and Contributions:

The Hale 200-inch Telescope quickly became a cornerstone of astronomical research, contributing to several significant discoveries:

- **Quasars:** One of the most notable discoveries made with the Hale Telescope was the identification of quasars, extremely bright and distant objects powered by supermassive black holes at the centers of galaxies. These discoveries provided insights into the early universe and the nature of black holes.

- **Structure of Galaxies:** The Hale Telescope was instrumental in mapping the structure and distribution of galaxies. Astronomers used the telescope to study the detailed properties of galaxies, including their shapes, sizes, and compositions, leading to a better understanding of galaxy formation and evolution.

- **Stellar Populations:** The telescope's powerful optics allowed astronomers to resolve individual stars in distant galaxies and star clusters. This enabled detailed studies of stellar populations, shedding light on the life cycles of stars and the chemical evolution of galaxies.

Technological Innovations:

The Hale 200-inch Telescope introduced several technological advancements that set new standards for astronomical instruments. The use of Pyrex glass for the mirror reduced thermal distortion, providing clearer images. The telescope's precise tracking system allowed for long exposure astrophotography, capturing faint and distant objects with unprecedented detail.

One of the key innovations was the implementation of a sophisticated control system that allowed astronomers to make fine adjustments to the telescope's position and focus. This system improved the accuracy and efficiency of observations, enabling more detailed and extensive studies of celestial objects.

Educational and Public Impact:

The Hale 200-inch Telescope not only advanced scientific knowledge but also played a significant role in education and public outreach. The Palomar Observatory became a center for astronomical research and training, attracting prominent scientists from around the world. It provided a platform for collaborative research and fostered the development of new techniques and technologies in astronomy.

Public interest in the Hale Telescope was high, and the observatory hosted numerous public viewing nights and educational programs. These events allowed people to experience the wonders of the universe firsthand and inspired many to pursue careers in science and astronomy. The telescope's achievements were widely publicized, helping to raise public awareness and appreciation of astronomical research.

The Legacy of the Hale 200-inch Telescope:

The Hale 200-inch Telescope remained the world's largest optical telescope until the 1990s, and its contributions to astronomy were immense. It played a crucial role in expanding our understanding of the universe and set new standards for the design and construction of large telescopes. The telescope's legacy is marked by the numerous discoveries it facilitated and the technological innovations it introduced.

The Hale Telescope demonstrated the importance of large-aperture telescopes for advancing astronomical research. Its success inspired the construction of even larger and more sophisticated telescopes, paving the way for future advancements in the field. The legacy of the Hale 200-inch Telescope continues to inspire new generations of astronomers and engineers, driving the quest to explore the universe further.

The Hale 200-inch Telescope represented a giant leap in astronomical observation and technology. Its groundbreaking discoveries and technological innovations expanded our knowledge of the cosmos and set new standards for astronomical research. The legacy of the Hale Telescope lives on, inspiring future generations and highlighting the importance of innovation and perseverance in the pursuit of scientific knowledge.

TELESCOPE 6: HORN ANTENNA (1959)

ECHOES OF THE BIG BANG

Introduction of the Horn Antenna:

In 1959, the Bell Telephone Laboratories in New Jersey built the Horn Antenna, originally designed for satellite communication experiments. However, this relatively simple radio telescope would soon become one of the most significant instruments in the history of astronomy, leading to the discovery of the cosmic microwave background radiation and providing critical evidence for the Big Bang theory.

The Construction of the Horn Antenna:

The Horn Antenna was designed as a large, horn-shaped structure, optimized for detecting radio waves. It measured 20 feet in length and was mounted on a rotating base, allowing it to be aimed at different parts of the sky. The antenna's unique design minimized interference and allowed for the precise measurement of weak radio signals.

Constructed primarily of aluminum, the Horn Antenna featured a smooth, conical shape that funneled radio waves into a receiver at its narrow end. This design reduced the noise from the antenna itself, making it highly sensitive to faint signals from space. The antenna's construction and positioning allowed for continuous tracking of celestial objects, essential for its role in detecting cosmic phenomena.

Discoveries and Contributions:

The Horn Antenna's most significant contribution came in 1965 when Arno Penzias and Robert Wilson used it to make a groundbreaking discovery:

- **Cosmic Microwave Background Radiation:** While conducting experiments with the Horn Antenna, Penzias and Wilson detected a persistent, faint noise that seemed to come from all directions. After ruling out various sources of interference, they realized they had discovered the cosmic microwave background radiation, the residual thermal radiation from the Big Bang. This discovery provided strong evidence for the Big Bang theory, fundamentally altering our understanding of the universe's origins.

The discovery of the cosmic microwave background radiation earned Penzias and Wilson the Nobel Prize in Physics in 1978. Their work with the Horn Antenna confirmed the predictions of the Big Bang theory and helped establish it as the leading explanation for the origin and evolution of the universe.

Technological Innovations:

The Horn Antenna introduced several technological innovations that made it exceptionally sensitive to faint radio signals. Its horn-shaped design minimized noise and interference, allowing for the precise detection of weak signals from space. The rotating mount enabled comprehensive sky surveys, making it possible to map the distribution of radio emissions across the sky.

Additionally, the antenna's receiver system was optimized for stability and accuracy, ensuring reliable measurements over extended periods. These technological advancements made the Horn Antenna a versatile and powerful instrument for radio astronomy, capable of detecting phenomena that had previously been beyond the reach of existing instruments.

Educational and Public Impact:

The discovery made with the Horn Antenna had a profound impact on both the scientific community and the general public. It provided concrete evidence for the Big Bang theory, helping to unify the fields of cosmology and particle physics. The finding sparked widespread interest in the origins of the universe and the nature of cosmic phenomena.

The Horn Antenna itself became a symbol of scientific ingenuity and the power of observational technology. Its role in one of the most significant discoveries of the 20th century highlighted the importance of radio astronomy and the potential for relatively simple instruments to make groundbreaking contributions to our understanding of the universe.

The Legacy of the Horn Antenna:

The Horn Antenna's legacy is marked by its pivotal role in the discovery of the cosmic microwave background radiation. This finding provided critical evidence for the Big Bang theory and transformed our understanding of the universe's origins and evolution. The antenna demonstrated the value of radio astronomy and the importance of precise, sensitive instruments in detecting cosmic phenomena.

The Horn Antenna continues to inspire new generations of astronomers and physicists, highlighting the potential for innovative technology to unlock the mysteries of the universe. Its contributions to cosmology remain a cornerstone of modern astrophysics, emphasizing the importance of observation and experimentation in advancing scientific knowledge.

The Horn Antenna, initially designed for satellite communication, became one of the most important instruments in the history of astronomy. Its discovery of the cosmic microwave background radiation provided critical evidence for the Big Bang theory and transformed our understanding of the universe. The legacy of the Horn Antenna endures, inspiring future generations of scientists and highlighting the power of technological innovation in the quest to explore the cosmos.

TELESCOPE 7:
ARECIBO OBSERVATORY (1963)

THE GIANT EAR

Introduction of the Arecibo Observatory:

In 1963, the Arecibo Observatory was inaugurated in Puerto Rico, becoming the world's largest and most powerful radio telescope at that time. With its enormous 1,000-foot (305-meter) diameter dish, Arecibo was designed to study the ionosphere, but its versatility allowed it to make significant contributions to various fields of astronomy, from planetary science to the search for extraterrestrial intelligence (SETI).

The Construction of the Arecibo Observatory:

The Arecibo Observatory was an engineering marvel. The massive dish was constructed in a natural limestone sinkhole, providing a stable foundation for its immense size. The dish was made up of perforated aluminum panels, supported by a network of steel cables and a series of towers that suspended a 900-ton platform above the dish.

This platform housed the telescope's receiver and various scientific instruments, which could be moved to different positions to focus on different parts of the sky. The unique design allowed Arecibo to track celestial objects by moving the receiver, rather than the entire dish, enabling it to observe a wide range of frequencies and conduct various types of scientific research.

Discoveries and Contributions:

The Arecibo Observatory quickly became a cornerstone of astronomical research, contributing to several significant discoveries:

- **Planetary Radar Mapping:** Arecibo's powerful radar system allowed astronomers to map the surfaces of planets and moons with unprecedented detail. Notable achievements included detailed radar maps of Venus, revealing surface features obscured by its thick atmosphere, and the discovery of ice at the poles of Mercury.

- **Pulsar Discovery:** In 1968, Jocelyn Bell Burnell and Antony Hewish discovered the first pulsar, a rapidly rotating neutron star emitting regular pulses of radio waves. Arecibo played a crucial role in the subsequent study of pulsars, leading to important insights into their properties and the nature of neutron stars.

- **Binary Pulsar and Gravitational Waves:** In 1974, Russell Hulse and Joseph Taylor used Arecibo to discover the first binary pulsar, two neutron stars orbiting each other. Their observations provided the first indirect evidence for the existence of gravitational waves, predicted by Einstein's theory of general relativity. This discovery earned them the Nobel Prize in Physics in 1993.

- **SETI and Extraterrestrial Intelligence:** Arecibo was instrumental in the search for extraterrestrial intelligence (SETI). In 1974, the observatory transmitted the Arecibo Message, a radio signal intended to demonstrate human technological achievement to potential extraterrestrial civilizations. Arecibo also conducted numerous SETI observations, scanning the sky for signals from intelligent sources.

Technological Innovations:

The Arecibo Observatory introduced several technological innovations that set new standards for radio astronomy. Its enormous size and powerful radar system allowed for high-resolution observations and detailed mapping of celestial objects. The observatory's unique design, with a movable receiver platform, enabled versatile and comprehensive sky surveys.

Arecibo's advanced instrumentation and data processing capabilities made it a versatile tool for various types of scientific research. It was equipped with sophisticated receivers, spectrometers, and data analysis systems, allowing astronomers to conduct detailed studies of radio emissions from a wide range of sources, from planets to distant galaxies.

Educational and Public Impact:

The Arecibo Observatory not only advanced scientific knowledge but also played a significant role in education and public outreach. The observatory became a center for training astronomers and engineers, providing hands-on experience with cutting-edge technology. Many notable scientists conducted research at Arecibo, contributing to its legacy of innovation and discovery.

Public interest in the Arecibo Observatory was high, and the facility hosted numerous educational programs and public tours. These initiatives helped to inspire future generations of scientists and raise public awareness and appreciation of astronomy. The observatory's iconic status and impressive achievements made it a symbol of scientific progress and exploration.

The Legacy of the Arecibo Observatory:

The Arecibo Observatory remained the world's largest and most powerful radio telescope for over five decades, making numerous contributions to astronomy and planetary science. Its discoveries have had a profound impact on our understanding of the universe, from the nature of neutron stars to the search for extraterrestrial intelligence.

In 2020, structural damage led to the collapse of the Arecibo Observatory's platform, marking the end of an era. However, its legacy lives on through the wealth of scientific data it collected and the inspiration it provided to generations of astronomers and the public. The observatory's achievements continue to be celebrated, highlighting the importance of perseverance, innovation, and the quest to explore the cosmos.

The Arecibo Observatory, with its giant ear to the cosmos, revolutionized radio astronomy and made groundbreaking discoveries that expanded our understanding of the universe. Its technological innovations, significant contributions to science, and role in education and public outreach cement its place in history as one of the most important astronomical instruments ever built. The legacy of Arecibo endures, inspiring future explorations and reminding us of the incredible potential of human ingenuity.

TELESCOPE 8:
WESTERBORK SYNTHESIS RADIO TELESCOPE (1970)

RADIO INTERFEROMETRY PIONEER

Introduction of the Westerbork Synthesis Radio Telescope:

In 1970, the Westerbork Synthesis Radio Telescope (WSRT) was inaugurated in the Netherlands. This powerful array of 14 radio antennas played a pivotal role in the development of radio interferometry and has made significant contributions to our understanding of the universe. WSRT's innovative design and advanced technology have made it a cornerstone of radio astronomy.

The Construction of the Westerbork Synthesis Radio Telescope:

The WSRT consists of 14 parabolic dish antennas, each 25 meters in diameter, aligned in a straight line over a distance of 2.7 kilometers. The antennas can be moved along railway tracks to create different configurations, allowing for high-resolution imaging and flexibility in observations. This design enabled astronomers to synthesize a large aperture, effectively creating a telescope with an enormous diameter.

Constructing the WSRT involved significant engineering challenges, particularly in aligning the antennas and developing the computer systems required to combine their signals. The innovative use of interferometry allowed the WSRT to achieve high-resolution images of celestial objects, making it one of the most advanced radio telescopes of its time.

Discoveries and Contributions:

The Westerbork Synthesis Radio Telescope has been instrumental in several key discoveries and advancements in radio astronomy:

- **Galactic Structure and Dynamics:** The WSRT has played a crucial role in mapping the structure and dynamics of the Milky Way and other galaxies. By studying the distribution of hydrogen gas, astronomers have gained insights into the formation and evolution of galaxies, as well as the processes driving star formation.

- **Pulsar Studies:** The WSRT has been used to discover and study pulsars, highly magnetized, rotating neutron stars that emit beams of electromagnetic radiation. These observations have provided valuable information about the properties and behavior of neutron stars.

- **Cosmic Magnetism:** The WSRT has contributed to our understanding of cosmic magnetism by mapping the magnetic fields of galaxies. These studies have helped to uncover the role of magnetic fields in galaxy formation and evolution.

- **Extragalactic Research:** The WSRT has been used to study distant galaxies and quasars, providing insights into the nature of these objects and the processes occurring in the early universe. Its high-resolution capabilities have allowed astronomers to resolve fine details in these distant objects.

Technological Innovations:

The Westerbork Synthesis Radio Telescope introduced several technological innovations that set new standards for radio astronomy. Its use of movable antennas and interferometry allowed for high-resolution imaging and flexibility in observations. The WSRT's advanced computer systems combined the signals from the individual antennas, creating detailed images of celestial objects.

The ability to reconfigure the array by moving the antennas along tracks provided astronomers with the versatility to conduct a wide range of observations, from large-scale surveys to detailed studies of specific objects. These technological advancements made the WSRT a powerful and versatile tool for radio astronomy.

Educational and Public Impact:

The Westerbork Synthesis Radio Telescope has had a significant impact on both the scientific community and the general public. As a major research facility, the WSRT has trained numerous astronomers and engineers, providing them with hands-on experience with advanced radio astronomy techniques. The observatory has been a hub for collaborative research, attracting scientists from around the world.

Public interest in the WSRT has been high, and the observatory has hosted numerous educational programs and public tours. These initiatives have helped to inspire future generations of scientists and raise public awareness and appreciation of radio astronomy. The WSRT's achievements have highlighted the importance of innovation and collaboration in scientific research.

The Legacy of the Westerbork Synthesis Radio Telescope:

The Westerbork Synthesis Radio Telescope has been a cornerstone of radio astronomy for over five decades, making numerous contributions to our understanding of the universe. Its technological innovations have set new standards for radio telescopes, influencing the design of subsequent arrays and observatories.

The WSRT's legacy is marked by its significant scientific achievements and its role in advancing the field of radio astronomy. The array continues to be a vital tool for astronomers, providing critical data for ongoing research and future discoveries. Its contributions to science and its impact on education and public outreach ensure that the WSRT will be remembered as one of the most important astronomical instruments of its time.

The Westerbork Synthesis Radio Telescope revolutionized radio astronomy with its innovative design and powerful capabilities. Its contributions to our understanding of galactic structure, pulsars, cosmic magnetism, and extragalactic research have been profound. The WSRT's legacy endures, inspiring future explorations and highlighting the incredible potential of human ingenuity in the quest to explore the cosmos.

TELESCOPE 9:
VERY LARGE ARRAY (1980)

RADIO VISIONARIES

Introduction of the Very Large Array:

In 1980, the Very Large Array (VLA) was inaugurated in the plains of San Agustin near Socorro, New Mexico. This impressive collection of 27 radio antennas, each 25 meters in diameter, arranged in a Y-shaped configuration, became one of the most powerful radio observatories in the world. The VLA has played a critical role in numerous astronomical discoveries, providing unprecedented insights into the universe.

The Construction of the Very Large Array:

The VLA's construction was an ambitious project led by the National Radio Astronomy Observatory (NRAO). Each of the 27 antennas could be moved along a set of railway tracks, allowing for different configurations that varied the array's resolution and sensitivity. When combined, the antennas could function as a single, massive radio telescope with an effective aperture of up to 36 kilometers.

The VLA's design was revolutionary, employing sophisticated computer algorithms to combine the signals from the individual antennas. This technique, known as interferometry, allowed the VLA to produce images with resolution and detail far surpassing those of single-dish radio telescopes. The flexibility of the array's configuration enabled astronomers to tailor their observations to specific scientific needs, making the VLA an incredibly versatile tool.

Discoveries and Contributions:

The Very Large Array has been instrumental in a wide range of astronomical discoveries and research projects:

- **Black Holes and Jets:** The VLA has provided detailed images of jets emitted by supermassive black holes at the centers of galaxies. These observations have shed light on the mechanisms powering these jets and their impact on galaxy evolution.

- **Star Formation:** The array has been used to study regions of star formation, revealing the processes by which stars and planetary systems are born. The VLA's ability to observe in multiple radio wavelengths has provided insights into the complex interactions of gas and dust in these regions.

Planetary Science: The VLA has contributed to our understanding of the planets within our solar system, particularly in studying the atmospheres of gas giants like Jupiter and Saturn. Its observations have helped to reveal the dynamics of planetary weather systems and the composition of their atmospheres.

- **Galactic Structure:** By mapping the distribution of hydrogen gas in the Milky Way, the VLA has helped to elucidate the structure of our galaxy, including the spiral arms and the central bulge. These studies have improved our understanding of galactic dynamics and the life cycle of stars.

Technological Innovations:

The VLA introduced several technological innovations that set new standards for radio astronomy. The use of interferometry to combine signals from multiple antennas allowed for unprecedented resolution and sensitivity. The array's ability to change configurations provided versatility, enabling observations of both wide fields and fine details.

The VLA's advanced computer systems processed vast amounts of data in real-time, generating high-resolution images of the radio sky. This capability revolutionized the field of radio astronomy, making it possible to conduct detailed surveys and targeted observations of specific astronomical objects.

Educational and Public Impact:

The Very Large Array has had a significant impact on both the scientific community and the general public. As a major research facility, the VLA has trained numerous astronomers and engineers, providing them with hands-on experience with cutting-edge technology. The observatory has been a hub for collaborative research, attracting scientists from around the world.

Public interest in the VLA has been high, partly due to its iconic status and its appearances in popular media, such as the film "Contact." The observatory offers tours and educational programs, allowing visitors to learn about radio astronomy and the array's contributions to science. These outreach efforts have inspired many to pursue careers in astronomy and related fields.

The Legacy of the Very Large Array:

The Very Large Array has been a cornerstone of radio astronomy for over four decades, making numerous contributions to our understanding of the universe. Its technological innovations have set new standards for radio telescopes, influencing the design of subsequent arrays and observatories.

The VLA's legacy is marked by its significant scientific achievements and its role in advancing the field of radio astronomy. The array continues to be a vital tool for astronomers, providing critical data for ongoing research and future discoveries. Its contributions to science and its impact on education and public outreach ensure that the VLA will be remembered as one of the most important astronomical instruments of its time.

The Very Large Array has revolutionized radio astronomy with its innovative design and powerful capabilities. Its contributions to our understanding of black holes, star formation, planetary science, and galactic structure have been profound. The VLA's legacy endures, inspiring future explorations and highlighting the incredible potential of human ingenuity in the quest to explore the cosmos.

TELESCOPE 10: HUBBLE SPACE TELESCOPE (1990)

EYES IN THE SKY

Introduction of the Hubble Space Telescope:

Launched in 1990, the Hubble Space Telescope (HST) has become one of the most significant and recognizable instruments in the history of astronomy. Orbiting above Earth's atmosphere, Hubble has provided unprecedented views of the universe, leading to numerous groundbreaking discoveries and transforming our understanding of the cosmos.

The Construction of the Hubble Space Telescope:

The Hubble Space Telescope was the result of decades of planning and collaboration between NASA and the European Space Agency (ESA). Named after the renowned astronomer Edwin Hubble, the telescope was designed to observe the universe in visible, ultraviolet, and near-infrared light. Its primary mirror, 2.4 meters (7.9 feet) in diameter, was meticulously polished to achieve the precise optical quality needed for clear, detailed observations.

Hubble's construction faced significant challenges, particularly in achieving the necessary optical precision and developing the instruments required for its diverse scientific missions. The telescope was equipped with a suite of advanced scientific instruments, including cameras, spectrographs, and fine guidance sensors, allowing it to conduct a wide range of astronomical observations.

Discoveries and Contributions:

The Hubble Space Telescope has made numerous contributions to our understanding of the universe, including:

- **Expansion of the Universe:** One of Hubble's most significant contributions has been the precise measurement of the rate of the universe's expansion, known as the Hubble constant. These measurements have provided critical insights into the age and size of the universe.

- **Deep Field Observations:** Hubble's deep field images, such as the Hubble Deep Field and the Hubble Ultra-Deep Field, have revealed thousands of previously unseen galaxies, providing a glimpse into the early universe and the formation of galaxies.

- **Dark Energy:** Observations of distant supernovae with Hubble have provided evidence for the existence of dark energy, a mysterious force driving the accelerated expansion of the universe. This discovery has profound implications for our understanding of cosmology and the fate of the universe.

- **Planetary Science:** Hubble has made detailed observations of planets within our solar system, including the discovery of new moons and the study of atmospheric phenomena. Its observations of exoplanets have provided insights into the composition and behavior of distant planetary systems.

Technological Innovations:

The Hubble Space Telescope introduced several technological innovations that set new standards for space-based observatories. Its precise optics and advanced scientific instruments have allowed for high-resolution imaging and spectroscopy across a wide range of wavelengths. Hubble's position above Earth's atmosphere eliminates the distortion caused by atmospheric turbulence, providing clearer and more detailed images than ground-based telescopes.

One of the key innovations was Hubble's ability to be serviced and upgraded by astronauts. Over the course of its mission, five servicing missions were conducted by the Space Shuttle program, allowing for repairs, upgrades, and the installation of new instruments. This capability has extended Hubble's operational life and ensured that it remains at the forefront of astronomical research.

Educational and Public Impact:

The Hubble Space Telescope has had a profound impact on both the scientific community and the general public. Its stunning images of distant galaxies, nebulae, and star clusters have captured the imagination of people around the world, raising public awareness and interest in astronomy.

Hubble's discoveries have been widely publicized, contributing to a greater understanding of the universe and our place within it. Educational programs and resources associated with Hubble have inspired countless students and educators, fostering a new generation of scientists and astronomers.

The Legacy of the Hubble Space Telescope:

The Hubble Space Telescope's legacy is marked by its numerous scientific achievements and its role in transforming our understanding of the universe. Its discoveries have provided critical insights into the nature of the cosmos, from the formation of galaxies to the behavior of dark energy.

Hubble's success has paved the way for future space-based observatories, such as the James Webb Space Telescope, which will build on its legacy and continue the quest to explore the universe. The telescope's ability to capture the beauty and complexity of the cosmos has left an indelible mark on both science and culture.

The Hubble Space Telescope has revolutionized our understanding of the universe with its groundbreaking discoveries and stunning images. Its contributions to cosmology, planetary science, and astrophysics have been profound, inspiring future explorations and highlighting the incredible potential of human ingenuity in the quest to explore the cosmos. The legacy of Hubble endures, reminding us of the power of observation and the endless possibilities of discovery.

TELESCOPE II: KECK 1 & 2 TELESCOPES (1993)

THE TWIN GIANTS

Introduction of the Keck 1 & 2 Telescopes:

In 1993, the W. M. Keck Observatory in Hawaii unveiled its first telescope, Keck 1, followed by its twin, Keck 2, in 1996. These twin giants, each with a 10-meter primary mirror composed of 36 hexagonal segments, became the most powerful optical telescopes in the world. Located on the summit of Mauna Kea, one of the best astronomical sites on Earth, the Keck telescopes have made significant contributions to a wide range of astronomical discoveries.

The Construction of the Keck Telescopes:

The construction of the Keck telescopes was a monumental engineering achievement. Each telescope's primary mirror consists of 36 hexagonal segments, which work together to form a single, large mirror with an effective diameter of 10 meters. These segments are precisely aligned using a system of actuators and sensors, ensuring that they function as a coherent optical surface.

The telescopes are housed in separate domes on the summit of Mauna Kea, where the thin, dry atmosphere provides optimal conditions for astronomical observations. The design includes advanced adaptive optics systems, which compensate for the blurring effects of Earth's atmosphere, allowing the telescopes to achieve near-diffraction-limited performance.

Discoveries and Contributions:

The Keck telescopes have been instrumental in numerous groundbreaking discoveries across various fields of astronomy:

- **Exoplanets:** The Keck telescopes have played a crucial role in the discovery and characterization of exoplanets. By measuring the tiny wobbles of stars caused by orbiting planets, astronomers have identified many new exoplanets and studied their properties, such as mass, composition, and atmospheric conditions.

- **Black Holes:** Keck's observations have provided detailed images and data on supermassive black holes at the centers of galaxies. These studies have revealed the dynamics of accretion disks, jet formation, and the interactions between black holes and their host galaxies.

- **Galactic Evolution:** The telescopes have been used to study the formation and evolution of galaxies across cosmic time. By observing distant galaxies, astronomers have gained insights into the processes that shape galaxies, such as mergers, star formation, and the influence of dark matter.

- **Stellar Populations:** The Keck telescopes have enabled detailed studies of stellar populations in various environments, from star clusters to the outskirts of the Milky Way. These observations have improved our understanding of stellar evolution, chemical enrichment, and the distribution of stars in the universe.

Technological Innovations:

The Keck telescopes introduced several technological innovations that set new standards for optical astronomy. The segmented primary mirror design allowed for the construction of larger mirrors than would be possible with a single piece of glass. This design also facilitated easier maintenance and repairs, as individual segments could be replaced or adjusted as needed.

The telescopes' advanced adaptive optics systems have revolutionized ground-based astronomy by correcting for atmospheric distortions in real-time. This technology allows the Keck telescopes to achieve image quality comparable to space-based telescopes, enabling high-resolution observations of faint and distant objects.

Educational and Public Impact:

The W. M. Keck Observatory has played a significant role in education and public outreach. The observatory collaborates with universities and research institutions, providing access to its state-of-the-art facilities for scientific research and training. Many students and early-career scientists have gained valuable experience and made important contributions to astronomy through their work at Keck.

Public interest in the Keck telescopes has been high, and the observatory offers various educational programs and public tours. These initiatives help to raise awareness and appreciation of astronomy, inspiring future generations of scientists and enthusiasts. The Keck telescopes' achievements are widely publicized, highlighting the importance of continued investment in astronomical research and technology.

The Legacy of the Keck Telescopes:

The Keck 1 and 2 telescopes have set new benchmarks for optical telescopes, demonstrating the potential of segmented mirror designs and advanced adaptive optics. Their contributions to our understanding of the universe are vast, spanning the study of exoplanets, black holes, galaxies, and stellar populations.

The legacy of the Keck telescopes is marked by their technological innovations and scientific achievements. They continue to be among the most productive and influential telescopes in the world, providing critical data for ongoing research and future discoveries. The success of the Keck telescopes has inspired the development of even larger and more sophisticated telescopes, such as the Thirty Meter Telescope (TMT), which aims to further expand our exploration of the cosmos.

The Keck 1 and 2 telescopes represent a giant leap forward in optical astronomy. Their groundbreaking discoveries and technological innovations have transformed our understanding of the universe and set new standards for future telescopes. The legacy of the Keck telescopes endures, inspiring future explorations and highlighting the incredible potential of human ingenuity in the quest to explore the cosmos.

TELESCOPE 12:
CHANDRA X-RAY OBSERVATORY (1999)

X-RAY VISION

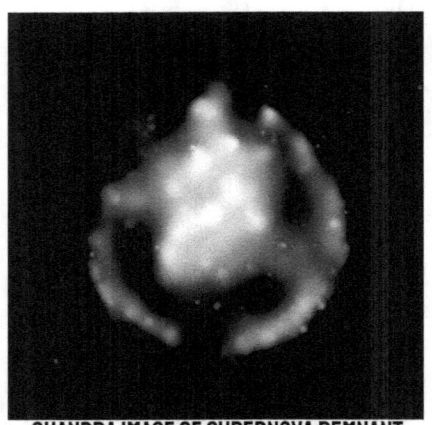

CHANDRA IMAGE OF SUPERNOVA REMNANT

Introduction of the Chandra X-ray Observatory:

Launched in 1999, the Chandra X-ray Observatory has provided astronomers with unparalleled views of the high-energy universe. Named after the Nobel Prize-winning astrophysicist Subrahmanyan Chandrasekhar, Chandra is designed to observe X-rays emitted by some of the most energetic and exotic phenomena in the cosmos, such as black holes, supernova remnants, and hot gas in galaxy clusters.

The Construction of the Chandra X-ray Observatory:

The construction of Chandra was a complex and ambitious project, led by NASA with significant contributions from other space agencies and research institutions. The observatory features a high-resolution X-ray telescope with nested mirrors coated with iridium, which are precisely shaped and aligned to focus X-rays onto advanced detectors.

Chandra's optics were designed to achieve unprecedented angular resolution, allowing it to produce sharp images of X-ray sources. The observatory is equipped with two main scientific instruments: the Advanced CCD Imaging

Spectrometer (ACIS) and the High Resolution Camera (HRC), both of which provide detailed imaging and spectroscopic capabilities.

Discoveries and Contributions:

Since its launch, the Chandra X-ray Observatory has made numerous groundbreaking discoveries across various fields of astronomy:

- **Black Holes:** Chandra has provided detailed images and spectra of black holes, revealing their properties and interactions with their surroundings. Observations of supermassive black holes at the centers of galaxies have shed light on their growth and the impact of their powerful jets on galaxy evolution.

- **Supernova Remnants:** The observatory has captured stunning images of supernova remnants, the remains of exploded stars. These observations have provided insights into the mechanisms of supernova explosions, the distribution of heavy elements, and the shock waves that shape these remnants.

- **Galaxy Clusters:** Chandra has been instrumental in studying the hot gas that fills galaxy clusters.

By mapping the distribution and temperature of this gas, astronomers have gained a better understanding of the formation and evolution of galaxy clusters, as well as the role of dark matter and dark energy.

- **Exotic Stars:** The observatory has detected and studied various exotic stars, such as neutron stars and white dwarfs. These observations have revealed the extreme conditions and physical processes occurring in these compact objects, including intense magnetic fields, rapid rotation, and the emission of powerful X-ray bursts.

Technological Innovations:

The Chandra X-ray Observatory introduced several technological innovations that set new standards for X-ray astronomy. Its advanced mirror design and precise alignment techniques have allowed for high-resolution imaging and accurate spectroscopy. The observatory's instruments are capable of detecting faint X-ray sources and resolving fine details, providing valuable data for a wide range of astrophysical studies.

Chandra's ability to operate in space, free from the interference of Earth's atmosphere, has been crucial for its success. The observatory's orbit takes it far above Earth's radiation belts, ensuring that it can conduct long-duration observations with minimal background noise. This has enabled Chandra to capture detailed images of some of the most distant and energetic phenomena in the universe.

Educational and Public Impact:

The Chandra X-ray Observatory has had a significant impact on both the scientific community and the general public. As a major research facility, Chandra has provided astronomers with a wealth of high-quality data, leading to numerous scientific publications and discoveries. The observatory has also played a key role in training the next generation of X-ray astronomers, offering opportunities for students and researchers to work with cutting-edge technology and data.

Public interest in Chandra has been high, thanks to the stunning images and exciting discoveries it has produced. The observatory's findings have been widely publicized, helping to raise awareness and appreciation of X-ray astronomy and the high-energy universe. Educational programs and outreach initiatives associated with Chandra have inspired countless students and educators, fostering a new generation of scientists and enthusiasts.

The Legacy of the Chandra X-ray Observatory:

The Chandra X-ray Observatory has set new benchmarks for X-ray astronomy, demonstrating the power of high-resolution imaging and spectroscopy in studying the universe's most energetic phenomena. Its contributions to our understanding of black holes, supernova remnants, galaxy clusters, and exotic stars have been profound, significantly advancing the field of high-energy astrophysics.

Chandra's success has paved the way for future X-ray observatories, such as the upcoming Lynx X-ray Observatory, which will build on its legacy and continue the quest to explore the high-energy universe. The observatory's ability to capture the beauty and complexity of X-ray sources has left an indelible mark on both science and culture.

The Chandra X-ray Observatory has revolutionized our understanding of the high-energy universe with its groundbreaking discoveries and stunning images. Its contributions to high-energy astrophysics have been profound, inspiring future explorations and highlighting the incredible potential of human ingenuity in the quest to explore the cosmos. The legacy of Chandra endures, reminding us of the power of observation and the endless possibilities of discovery.

TELESCOPE 13:
SPITZER SPACE TELESCOPE (2003)

INFRARED EXPLORER

Introduction of the Spitzer Space Telescope:

Launched in 2003, the Spitzer Space Telescope has provided astronomers with unprecedented views of the universe in infrared light. Named after the theoretical physicist Lyman Spitzer, who first proposed the idea of space telescopes, Spitzer has made significant contributions to our understanding of star formation, planetary systems, and the distant universe.

The Construction of the Spitzer Space Telescope:

The Spitzer Space Telescope was part of NASA's Great Observatories program, which also includes the Hubble Space Telescope, the Compton Gamma Ray Observatory, and the Chandra X-ray Observatory. Spitzer's primary mirror, measuring 85 centimeters (33.5 inches) in diameter, was designed to be cooled to just a few degrees above absolute zero, reducing the telescope's own infrared emissions and allowing it to detect faint heat signatures from distant celestial objects.

Spitzer was equipped with three main scientific instruments: the Infrared Array Camera (IRAC), the Infrared Spectrograph (IRS), and the Multiband

Imaging Photometer for Spitzer (MIPS). These instruments provided a wide range of imaging and spectroscopic capabilities, enabling detailed studies of astronomical phenomena across the infrared spectrum.

Discoveries and Contributions:

The Spitzer Space Telescope has made numerous groundbreaking discoveries across various fields of astronomy:

- **Star Formation:** Spitzer's infrared capabilities allowed astronomers to peer through dense clouds of gas and dust to observe the process of star formation. The telescope captured stunning images of star-forming regions, revealing the intricate structures and interactions that give rise to new stars and planetary systems.

- **Exoplanets:** Spitzer played a crucial role in the study of exoplanets, detecting the infrared signatures of planets orbiting other stars. The telescope was able to measure the temperatures, compositions, and atmospheres of exoplanets, providing valuable insights into their properties and potential habitability.

- **Galactic Evolution:** By observing distant galaxies in infrared light, Spitzer provided a glimpse into the early universe and the processes that shaped galaxy formation and evolution. The telescope's deep field observations revealed thousands of previously unseen galaxies, helping to map the growth and distribution of galaxies over cosmic time.

- **Dust and Debris Disks:** Spitzer detected and studied the dusty debris disks around young stars, which are thought to be the sites of planet formation. These observations provided evidence for the presence of young planetary systems and the processes that govern their development.

Technological Innovations:

The Spitzer Space Telescope introduced several technological innovations that set new standards for infrared astronomy. Its advanced cooling system, which included a liquid helium cryostat, allowed the telescope to operate at very low temperatures, minimizing its own infrared emissions and maximizing its sensitivity to faint cosmic signals.

Spitzer's instruments were designed to cover a wide range of infrared wavelengths, from the near-infrared to the far-infrared. This versatility enabled the telescope to conduct a broad array of scientific investigations, from detailed studies of individual stars and planets to large-scale surveys of galaxies and the interstellar medium.

Educational and Public Impact:

The Spitzer Space Telescope has had a significant impact on both the scientific community and the general public. As a major research facility, Spitzer provided astronomers with a wealth of high-quality data, leading to numerous scientific publications and discoveries. The observatory also played a key role in training the next generation of infrared astronomers, offering opportunities for students and researchers to work with cutting-edge technology and data.

Public interest in Spitzer has been high, thanks to the stunning images and exciting discoveries it has produced. The telescope's findings have been widely publicized, helping to raise awareness and appreciation of infrared astronomy and the hidden universe. Educational programs and outreach initiatives associated with Spitzer have inspired countless students and educators, fostering a new generation of scientists and enthusiasts.

The Legacy of the Spitzer Space Telescope:

The Spitzer Space Telescope's legacy is marked by its numerous scientific achievements and its role in advancing the field of infrared astronomy. Its contributions to our understanding of star formation, exoplanets, galactic evolution, and planetary systems have been profound, significantly advancing our knowledge of the universe.

Although Spitzer's primary mission ended in 2009 when it exhausted its supply of liquid helium coolant, the telescope continued to operate in a "warm" mission phase, using its remaining instruments to conduct valuable observations. The data collected by Spitzer will continue to be a rich resource for astronomers for years to come.

The Spitzer Space Telescope has revolutionized our understanding of the infrared universe with its groundbreaking discoveries and stunning images. Its contributions to star formation, exoplanet research, galactic evolution, and planetary systems have been profound, inspiring future explorations and highlighting the incredible potential of human ingenuity in the quest to explore the cosmos. The legacy of Spitzer endures, reminding us of the power of infrared observation and the endless possibilities of discovery.

TELESCOPE 14:
ATACAMA LARGE MILLIMETER ARRAY (ALMA, 2011)

THE COSMIC EYE

Introduction of the Atacama Large Millimeter Array:

In 2011, the Atacama Large Millimeter Array (ALMA) was inaugurated in the high-altitude Atacama Desert of northern Chile. ALMA, an international collaboration between Europe, North America, East Asia, and Chile, is the world's most powerful telescope for observing the universe at millimeter and submillimeter wavelengths. ALMA's ability to peer into the cold, dusty regions of space has revolutionized our understanding of the cosmos.

The Construction of ALMA:

ALMA consists of 66 high-precision antennas, each up to 12 meters in diameter, spread across the Chajnantor Plateau at an altitude of 5,000 meters (16,400 feet). The array can be reconfigured to cover distances from 150 meters to 16 kilometers, allowing it to achieve both wide-field views and incredibly high-resolution imaging.

The construction of ALMA involved significant logistical and engineering challenges due to its remote location and extreme altitude. The antennas and their components were transported to the site and assembled with great precision. The array's design includes state-of-the-art receivers and electronics that enable it to capture faint signals from the coldest and most distant regions of the universe.

Discoveries and Contributions:

ALMA has made numerous groundbreaking discoveries across various fields of astronomy, including:

- **Star Formation:** ALMA's ability to observe in millimeter wavelengths allows it to see through the dense clouds of gas and dust that obscure star-forming regions in optical light. This has provided detailed images of the processes involved in the birth of stars and planetary systems, including the formation of protoplanetary disks around young stars.

- **Galactic Evolution:** By observing distant galaxies, ALMA has provided insights into the early stages of galaxy formation and evolution. Its observations of cold gas and dust have revealed the conditions that led to the growth and development of galaxies over cosmic time.

- **Black Holes and Active Galactic Nuclei:** ALMA has been used to study the environments around supermassive black holes at the centers of galaxies. These observations have helped to understand the interactions between black holes and their host galaxies, including the mechanisms that drive the emission of powerful jets and outflows.

- **Planetary Formation:** ALMA has provided detailed images of protoplanetary disks, showing the structures and gaps where planets are forming. These observations have offered new insights into the processes that lead to planet formation and the diversity of planetary systems.

Technological Innovations:

ALMA introduced several technological innovations that set new standards for millimeter and submillimeter astronomy. Its large number of antennas and the ability to reconfigure the array provide unparalleled resolution and sensitivity. The array's advanced receivers are capable of detecting faint signals from the coldest regions of space, allowing astronomers to study the universe in unprecedented detail.

One of ALMA's key innovations is its use of interferometry to combine the signals from its individual antennas. This technique allows ALMA to achieve the resolution of a telescope with an aperture as large as the distance between its farthest antennas. The array's sophisticated data processing systems convert these signals into high-resolution images, providing detailed views of the universe at millimeter wavelengths.

Educational and Public Impact:

The Atacama Large Millimeter Array has had a significant impact on both the scientific community and the general public. As a major research facility, ALMA has provided astronomers with a wealth of high-quality data, leading to numerous scientific publications and discoveries. The observatory has also played a key role in training the next generation of astronomers, offering opportunities for students and researchers to work with cutting-edge technology and data.

Public interest in ALMA has been high, thanks to the stunning images and exciting discoveries it has produced. The observatory's findings have been widely publicized, helping to raise awareness and appreciation of millimeter and submillimeter astronomy. Educational programs and outreach initiatives associated with ALMA have inspired countless students and educators, fostering a new generation of scientists and enthusiasts.

The Legacy of ALMA:

The Atacama Large Millimeter Array's legacy is marked by its numerous scientific achievements and its role in advancing the field of millimeter and submillimeter astronomy. Its contributions to our understanding of star formation, galactic evolution, black holes, and planetary formation have been profound, significantly advancing our knowledge of the universe.

ALMA's success has paved the way for future observatories, such as the upcoming Next Generation Very Large Array (ngVLA), which will build on its legacy and continue the quest to explore the universe at millimeter wavelengths. The data collected by ALMA will continue to be a rich resource for astronomers for years to come, providing valuable insights into the cosmos.

The Atacama Large Millimeter Array has revolutionized our understanding of the cold and dusty universe with its groundbreaking discoveries and stunning images. Its contributions to star formation, galactic evolution, black holes, and planetary formation have been profound, inspiring future explorations and highlighting the incredible potential of human ingenuity in the quest to explore the cosmos. The legacy of ALMA endures, reminding us of the power of millimeter and submillimeter observations and the endless possibilities of discovery.

TELESCOPE 15:
JAMES WEBB SPACE TELESCOPE (2021)

A NEW VISION

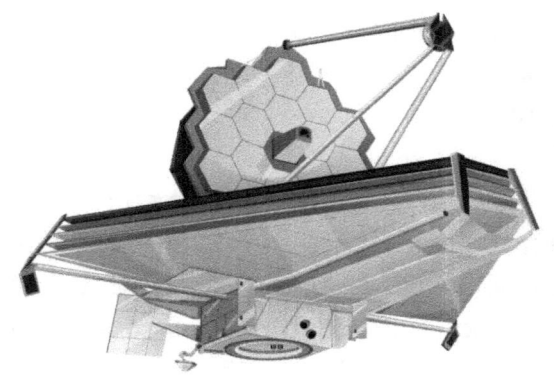

Introduction of the James Webb Space Telescope:

Launched in 2021, the James Webb Space Telescope (JWST) represents a new era in space-based astronomy. As the successor to the Hubble Space Telescope, JWST is designed to observe the universe in infrared light, providing unparalleled insights into the formation of stars and galaxies, the atmospheres of exoplanets, and the early universe. Named after James E. Webb, NASA's second administrator, the telescope is a collaborative project involving NASA, the European Space Agency (ESA), and the Canadian Space Agency (CSA).

The Construction of the James Webb Space Telescope:

The construction of JWST was a monumental engineering achievement that spanned over two decades. The telescope features a 6.5-meter (21.3-foot) primary mirror composed of 18 hexagonal segments made of beryllium and coated with gold. This large mirror is designed to capture more light than any previous space telescope, allowing JWST to observe faint and distant objects in the infrared spectrum.

One of the key innovations of JWST is its sunshield, a five-layer structure the size of a tennis court, which protects the telescope from the heat of the Sun and allows it to operate at extremely low temperatures. The observatory's instruments include the Near Infrared Camera (NIRCam), the Near Infrared Spectrograph (NIRSpec), the Mid-Infrared Instrument (MIRI), and the Fine Guidance Sensor/Near InfraRed Imager and Slitless Spectrograph (FGS/NIRISS), each designed to perform a wide range of scientific tasks.

Discoveries and Contributions:

Although still early in its mission, the James Webb Space Telescope is expected to make numerous groundbreaking discoveries:

- **Early Universe:** JWST is designed to look back in time to the formation of the first galaxies and stars. By observing these ancient light sources, the telescope will provide insights into the processes that shaped the early universe and the formation of complex structures.

- **Star and Planet Formation:** JWST's infrared capabilities will allow it to peer through dense clouds of gas and dust to study the birth of stars and planetary systems. These observations will help to understand the mechanisms of star formation and the early stages of planet development.

- **Exoplanet Atmospheres:** One of JWST's most exciting prospects is the detailed study of exoplanet atmospheres. By analyzing the light passing through these atmospheres, the telescope will be able to detect the presence of molecules such as water, carbon dioxide, and methane, providing clues about the potential habitability of these distant worlds.

- **Galactic Evolution:** JWST will also study the formation and evolution of galaxies across cosmic time. Its high-resolution images and spectra will reveal the interactions, mergers, and growth of galaxies, offering a deeper understanding of the dynamic processes that drive galactic evolution.

Technological Innovations:

The James Webb Space Telescope introduced several technological innovations that set new standards for space-based astronomy. The segmented primary mirror allows for a larger aperture than a single-piece mirror, enhancing the telescope's light-gathering capability. The use of beryllium and gold coatings ensures optimal performance in the infrared spectrum.

JWST's sunshield is another groundbreaking feature, providing the necessary thermal protection for the telescope's sensitive instruments. The sophisticated cryocooler system for the MIRI instrument allows it to operate at temperatures close to absolute zero, essential for detecting faint mid-infrared signals. The observatory's instruments are equipped with advanced detectors and spectrometers, enabling high-resolution imaging and detailed spectroscopic analysis.

Educational and Public Impact:

The James Webb Space Telescope has generated immense interest and excitement among both the scientific community and the general public. As a flagship mission for NASA and its international partners, JWST has been the focus of extensive outreach and educational programs. These initiatives aim to inspire the next generation of scientists and engineers, fostering a greater understanding of space science and the universe.

Public engagement with JWST has been high, with widespread media coverage and numerous events celebrating its launch and early observations. The stunning images and groundbreaking discoveries anticipated from JWST are expected to captivate the public imagination and highlight the importance of space exploration.

The Legacy of the James Webb Space Telescope:

The James Webb Space Telescope is poised to become one of the most important astronomical observatories in history. Its advanced capabilities and ambitious scientific goals promise to transform our understanding of the universe, building on the legacy of Hubble and other great observatories.

JWST's success will pave the way for future space telescopes, demonstrating the potential of advanced technologies and international collaboration in achieving scientific breakthroughs. The data collected by JWST will be a valuable resource for astronomers for decades, contributing to a wide range of scientific fields and enhancing our knowledge of the cosmos.

The James Webb Space Telescope represents a new vision for space-based astronomy, offering unprecedented capabilities and promising groundbreaking discoveries. Its contributions to our understanding of the early universe, star and planet formation, exoplanet atmospheres, and galactic evolution will be profound, inspiring future explorations and highlighting the incredible potential of human ingenuity in the quest to explore the cosmos. The legacy of JWST will endure, reminding us of the power of observation and the endless possibilities of discovery.

Conclusion

The journey through the history of these 15 remarkable telescopes has taken us from Galileo's humble tube to the cutting-edge James Webb Space Telescope, highlighting the extraordinary advancements in our understanding of the universe. Each telescope has played a crucial role in expanding our knowledge, pushing the boundaries of technology, and inspiring generations of scientists and enthusiasts.

Galileo's Telescope began this journey by challenging the geocentric view and opening humanity's eyes to a universe filled with moons, stars, and planets that defied conventional wisdom. This spirit of curiosity and innovation continued with the Yerkes 40-inch Refractor and the Mount Wilson 60-inch Reflector, which provided deeper insights into the structure and dynamics of our galaxy.

The Hooker 100-inch Telescope revealed the expanding universe, fundamentally changing our understanding of the cosmos and our place within it. The Hale 200-inch Telescope further pushed the limits of optical astronomy, while the Horn Antenna's detection of cosmic microwave background radiation provided critical evidence for the Big Bang theory.

Radio telescopes like the Arecibo Observatory and the Westerbork Synthesis Radio Telescope pioneered new ways of listening to the universe, unveiling the hidden structures and processes within it. The Very Large Array, with its innovative design, offered detailed radio images of celestial objects, enhancing our comprehension of galactic and extragalactic phenomena.

Space-based telescopes such as the Hubble Space Telescope revolutionized our view with stunning clarity and depth, capturing iconic images that have captivated the public and scientists alike. The Keck 1 & 2 Telescopes, with their advanced optics and location on Mauna Kea, have provided unparalleled observations of distant galaxies and exoplanets.

The Chandra X-ray Observatory offered a unique window into the high-energy universe, revealing the violent and energetic processes that shape celestial objects. The Spitzer Space Telescope explored the infrared universe, uncovering the secrets of star and planet formation in the cold and dusty regions of space.

The Atacama Large Millimeter Array (ALMA) has provided detailed views of the cold universe, revealing the processes of star formation and galaxy evolution with unprecedented precision. Finally, the James Webb Space Telescope, with its advanced capabilities and ambitious scientific goals, promises to further transform our understanding of the universe, building on the legacy of its predecessors.

As we reflect on the achievements of these extraordinary telescopes, we are reminded of the relentless human drive to explore, understand, and marvel at the cosmos. Each telescope, a testament to human ingenuity and perseverance, has contributed to a greater understanding of our universe and our place within it.

The legacy of these telescopes endures, inspiring future explorations and discoveries. They remind us that the quest to explore the cosmos is a journey without end, driven by our insatiable curiosity and the desire to uncover the mysteries of the universe. As we look to the future, new telescopes and observatories will continue this grand tradition, building on the foundations laid by these pioneering instruments and taking us ever closer to the stars.

References

- **Discoveries and Opinions of Galileo** by Stillman Drake. Anchor Books, 1957.

- **Edwin Hubble: Mariner of the Nebulae** by Gale E. Christianson. University of Chicago Press, 1995.

- **Big Bang: The Origin of the Universe** by Simon Singh. Harper Perennial, 2005.

- **Yerkes Observatory, 1892-1950: A Century of Astronomy** by Donald E. Osterbrock. University of Chicago Press, 1997.

- **The Mount Wilson Observatory: Breaking the Code of Cosmic Evolution** by Allan Sandage. Cambridge University Press, 2004.

- **The Space Telescope: A Study of NASA, Science, Technology, and Politics** by Robert W. Smith. Cambridge University Press, 1993.

- **The Cosmic Microwave Background Radiation** by Arno Penzias and Robert Wilson. Nobel Lecture, 1978.

- **VLBI and the VLBA: Evolution of High-Resolution Radio Interferometry by** Kenneth I. Kellermann and Jim Cordes. American Institute of Physics, 1991.

- **More Things in the Heavens: How Infrared Astronomy Is Expanding Our View of the Universe by** Michael Werner et al. Princeton University Press, 2019.

- **The James Webb Space Telescope** by Jonathan P. Gardner et al. Springer, 2006.